FSC
www.fsc.org
MIX
Papier aus ver-
antwortungsvollen
Quellen
Paper from
responsible sources
FSC® C105338

AF293566

Modelling of Hysteresis

Roland Büchi

Bibliografische Information der Deutschen Nationalbibliothek
Die Deutsche Nationalbibliothek verzeichnet diese Publikation in
der Deutschen Nationalbibliografie; detaillierte bibliografische
Daten sind im Internet über www.dnb.de abrufbar.

Impressum
© 2022, Roland Büchi
Herstellung und Verlag: BoD – Books on Demand,
Norderstedt
ISBN: 978-3-7562-9369-8

Modelling of Hysteresis

Modelling of Hysteresis

In order to model a real system, differential equations are usually derived, which are based on physical principles. When modelling a characteristic hysteresis, the mathematical equations are also based on the physics underlying hysteresis, but they are nonlinear effects. Therefore, a different approach is taken in the modelling. A measurement is used as the basis for modelling. This means that a specific measurement curve that determines the parameters must first be available. The advantage of this method is that it can be applied to many technical fields. The method presented here can be verified using the example of piezo elements, but it can also be applied to the magnetization characteristic of ferromagnetic materials. The methods of Preisach and Mayergoyz are presented and explained in this booklet. Examples are calculated with Matlab. The Matlab source code is printed at the end.

1.   Introduction to hysteresis

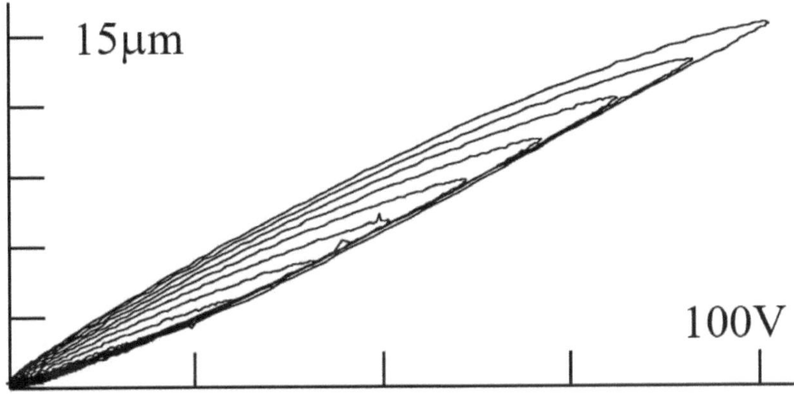

Fig.1: Displacement-/Voltage Diagram of a eines Piezo-
element

Figure 1 shows the displacement/voltage diagram of a multilayer piezo element measured using laser interferometry. It shows that at an applied voltage of 50V, the expansion can be between 6μm and 10μm depending on the history. The following shows how it is possible to simulate such a hysteresis using a mathematical model. This is treated as static non-linearity. Static means that the dynamics of the input signal have no influence on the properties and behavior of the hysteresis. This assumption is correct for quasi-static motions, i.e. for motions whose

static hystereses can be divided into two different classes [Mayergoyz 91]: there are those with local memory (Fig. 2a) and those with non-local memory (Fig. 2b).

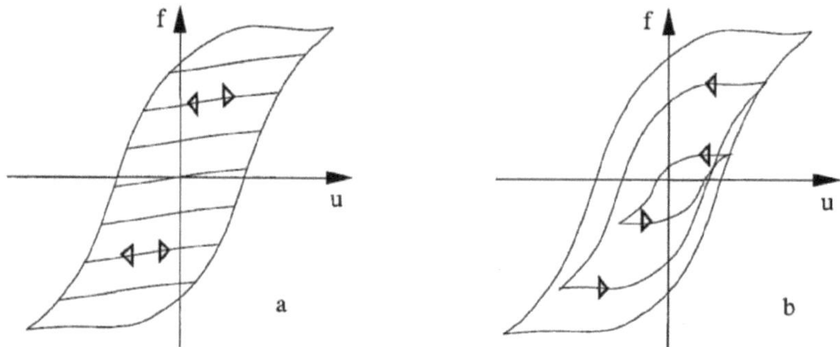

Fig.2:   Hysteresis with local (a) and non-local memory (b)

Local memory hystereses are characterized as follows: the output signal $f(t)$; $t = t0$ and the input signal $u(t)$; $t > t0$ together determine the output signal $f(t)$; $t > t0$. The output signal $f(t)$ is therefore not only dependent from the input signal, but also from a so-called local memory. It is therefore also dependent on the output signal $f(t)$; $t = t0$. For hystereses with non-local memory, the behavior becomes even a bit more complicated; here the output signal is dependent on the one hand on the current input signal $f(t)$ and on the other hand also on past input extreme values, that is to say on local maxima and minima. Simple measurements show that local maxima and minima must be stored for piezo elements. Thus, they belong to the second category, with non-local memory hysteresis. In the following, a methodology is developed for the mathematical modelling of such hysteresis.

2. Definition of the Preisach hysteresis-model

The origin of the hysteresis model discussed here goes back to F. Preisach [Preisach 35]. There, an infinitely large set of elementary hysteresis operators $\gamma_{\alpha\beta}$ (Fig. 3) is given.

These can be understood as a rectangular loop, with α representing the rising edge and ß the falling edge. It is assumed that $\alpha \geq$ ß. This is justified from a physical point of view. The outputs of these operators can only have the value +1 or -1. All operators $\gamma_{\alpha\beta}$ are weighted with a function $\mu(\alpha,ß)$. This weight function represents the hysteresis curve determined from measured values and is dealt with further below. The output function is calculated with the integral of those weighted elementary hysteresis operators.

The output function is calculated using formula (1)

$$f(t) = \iint_{\alpha \geq \beta} \mu(\alpha, \beta) \cdot \gamma_{\alpha,\beta} \cdot u(t)\, d\alpha\, d\beta$$

$$\tag{1}$$

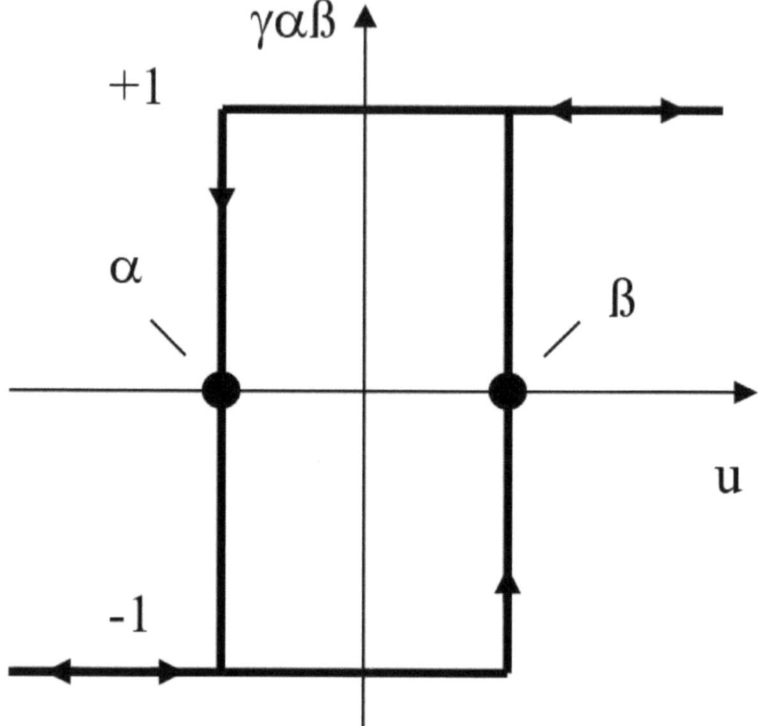

Fig. 3: Elementary hysteresis operator γαß

3. Storage and erase of local minima and maxima

The mathematical description of Preisach's hysteresis model is simplified by the geometric interpretation. This is based on the following: there is a direct connection between the hysteresis operators γαß and the points (ß/α) of a half-plane α ≥ ß. Each point of this half-plane corresponds exactly to the hysteresis operator with α and ß as "upward" and "downward switching point". Finally, the weighting function µ(α,ß) states how strongly the hysteresis operator γαß is weighted. Formula (1) shows that the output of the hysteresis function corresponds to a double integration with α and ß. In the geometric interpretation, the maximum value of the output f corresponds to the area spanned under the half-plane α ≥ ß. During the movement, a distinction is made as to whether the derivation of the input signal, in the case of the piezo elements the input voltage, is positive (up) or negative (down).

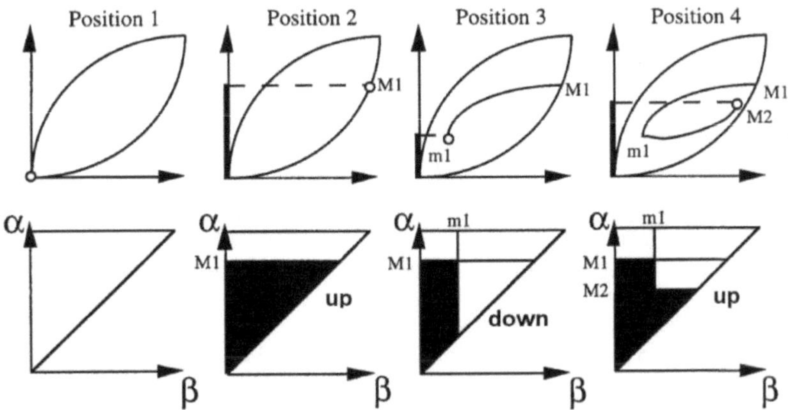

Fig. 4: Storage of local minima and maxima

The local minima and maxima now sit on the points where the derivatives of the input signal are zero: minima occur when changing from negative to positive, maxima when changing from positive to negative. Since multilayer piezo elements are only controlled with positive voltages, the relevant area of the half-plane $\alpha \geq \beta$ is only $\alpha \geq 0$ and $\beta \geq 0$. To describe this, a possible scenario according to Fig. 4 is played through:

Position 1 is the initial position. There is no voltage applied to the piezo element, the expansion is zero. At position 2, the input signal is steadily increased until a point M1 is reached. The content of the spanned area corresponds to the output of the piezo hysteresis. It should be noted that during this phase (up) it is terminated by a straight line which goes through M1 and is parallel to the β-axis. With rising input signals, the α-value decides when the hysteresis operators change to positive. For position 3, the input signal is steadily reduced again, starting from M1, until M1 is reached. When examining the $\alpha \geq \beta$ diagram, it can be determined that during this phase (down) the spanned area is now closed by a straight line through ml parallel to the α- axis. Analogously, the β value decides when the hysteresis operators change to negative. As can also be seen from the $\alpha \geq \beta$ diagram, the local maximum Ml is stored. At position 4 the input signal is again steadily increased until a point M2 is reached. The spanned area is again delimited by a straight line parallel to the β-axis. It is noticeable that the storage of the maxima and minima has led to a "stairways function" which contains all the necessary information.

It was shown above how local minima and maxima are stored. The question immediately arises as to whether it is necessary to save all past minima and maxima, or whether these can also be erased under certain circumstances. In order to get clarity about this, Fig. 5 should be considered in more detail.

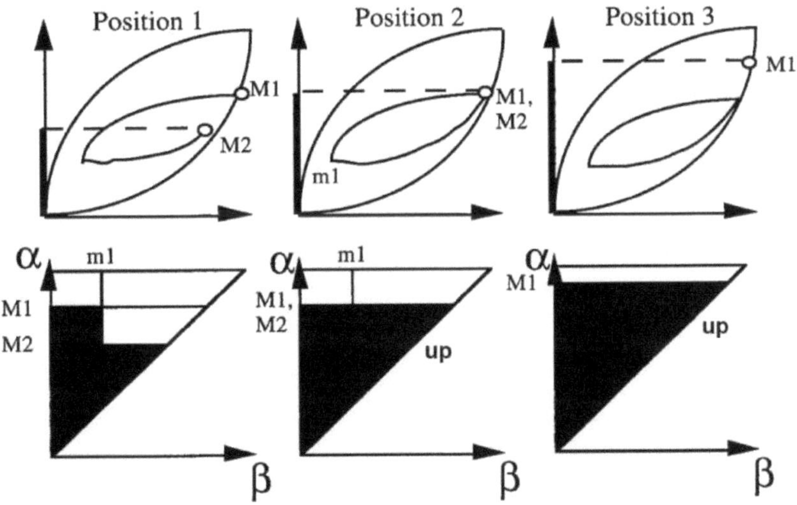

Fig. 5: Erase of local maxima and minima

Position 1 is exactly the same as position 4 of figure 4. If the input signal is steadily increased from M2 until M1 is reached, then M1 = M2 (position 2). If it is now further increased, as position 3 shows, then the stored loop, which contains the local maximum Ml and the local minimum ml, disappears. This means that if an old maximum is exceeded or an old minimum is undershot, the stored loop disappears. This can also be seen from the $\alpha \geq \beta$ diagram: while position 1 still has a

"stairways function", this is no longer available in position 3.
Only the new local maximum M1 is stored.

4.   The weight function μ(α,ß)

Up to this point, a homogeneously distributed $\alpha \geq ß$ plane
has been assumed. This is of course not the case with a real
piezohysteresis. This chapter shows how the weight function
μ(α,ß) can be derived from a measurement.

$$\mu(\alpha,\beta) = \frac{d^2 F(\alpha,\beta)}{d\alpha\, d\beta}$$

(2)

Given is a simple hysteresis as shown in figure 6. This
hysteresis can be described with very few values of an area
function. It is used as a didactic example to calculate the
weight function using formula 2.

In order to find the weight function, it must be measured in
the same way as the piezo hysteresis in figure 1. It is therefore
started at the point (0/0). First, the input is steadily increased
until point (4/4) is reached. Then it is reduced again to the
starting point (0/0).

Since discrete points are measured here, the $\alpha \geq \beta$ plane is divided into small squares. The output values of the largest loop (4, 3.5, 3, 2, 0) are now written in the top row of the area function $F(\alpha,\beta)$. In this way, all loops are measured and their values are written into the corresponding row (Fig. 7).

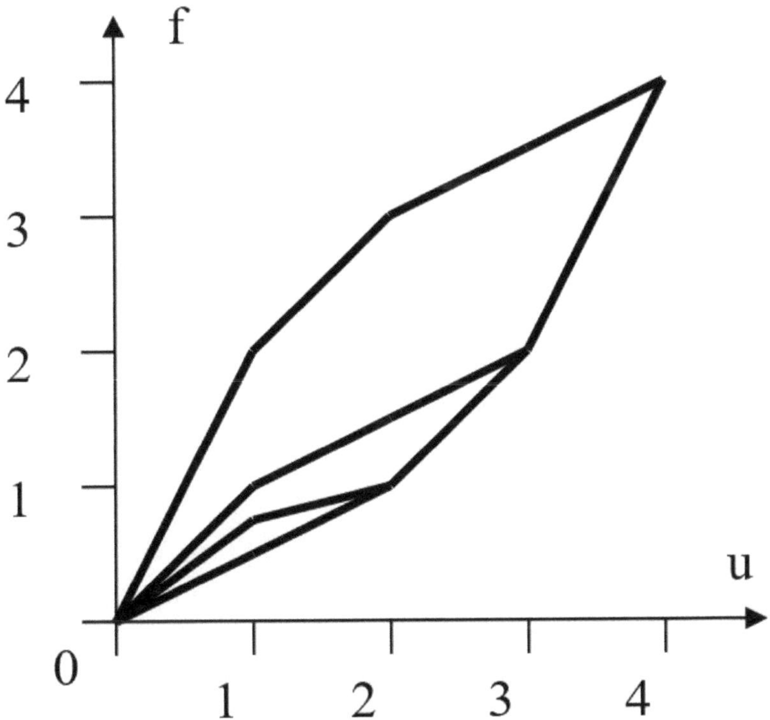

Fig. 6: Example of a hysteresis

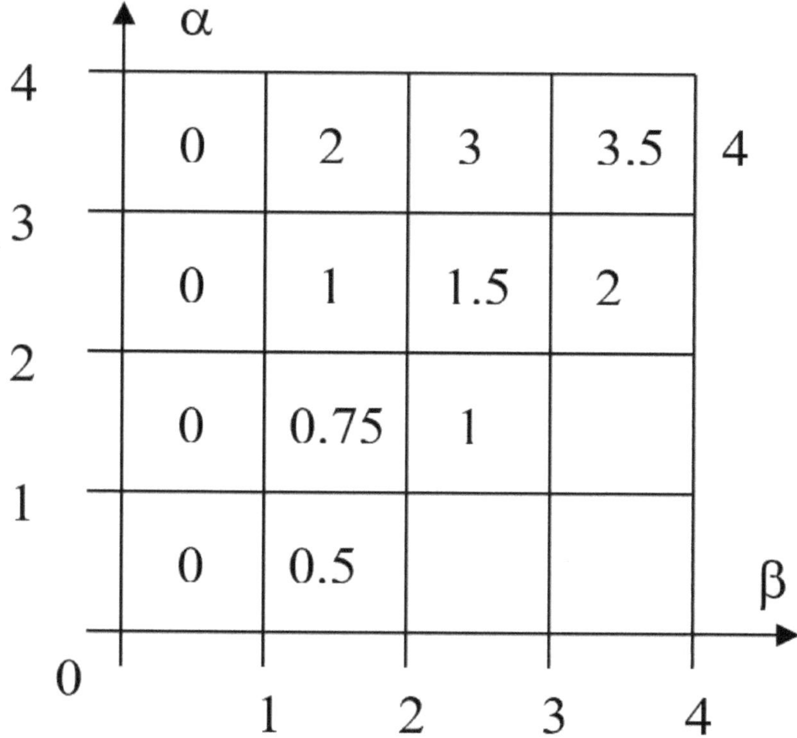

Fig. 7: F(α, ß)

In order to obtain the weight function $\mu(\alpha,ß)$, the area function F(α,ß) now has to be derived after α and ß, as shown in Fig. 8. You can easily understand this in the example by first calculating the horizontal differences for F(α,ß) and writing them in a diagram and then calculating the vertical differences and write them into this diagram.

Since the values obtained are not all of the same size, the weight distribution is not homogeneous. So not all hysteresis operators influence the output to the same extent. Thus, a piezo element can now be measured correctly and the corresponding weight function can be found. Local maxima and minima can also be saved via the software, which determine the "stairways function" of the output.

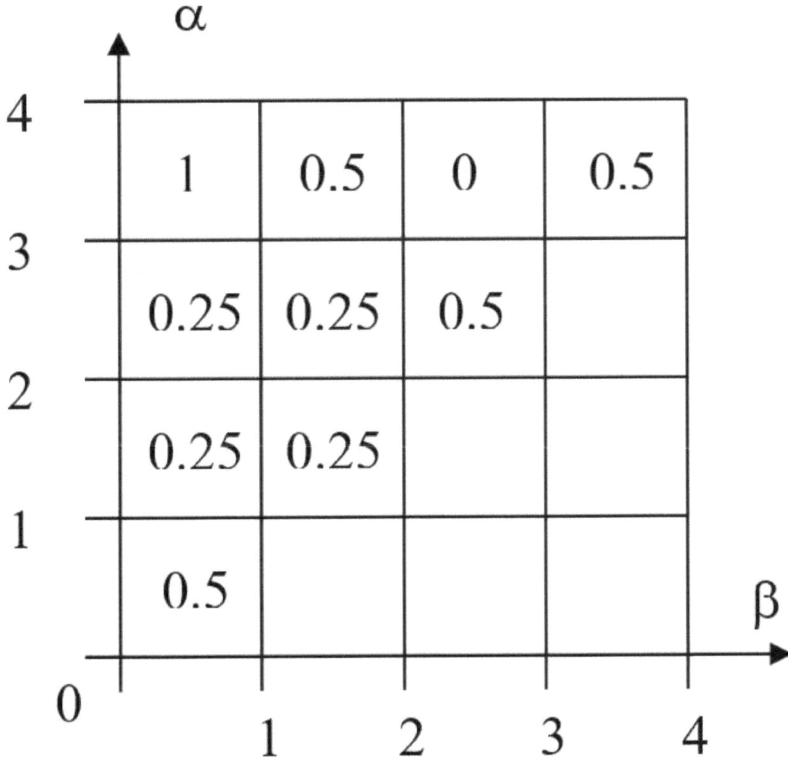

Fig. 8: μ (α , ß)

5. Implementation and verification

What it still missing is a way to implement this as a mathematical model. For this purpose, formulas were derived from [Mayergoyz 91], which are based on the area function F(α,ß). These were slightly adapted for the 'multilayer piezo element' problem (3).

$$
\begin{aligned}
f &= \sum_{k=1}^{n-1}\left[F(M_k, m_k) - F(M_k, m_{k-1})\right] \\
&\quad + \left[F(M_n, u) - F(M_n, m_{n-1})\right] \quad \textbf{down}
\end{aligned}
$$

$$
\begin{aligned}
f &= \sum_{k=1}^{n-1}\left[F(M_k, m_k) - F(M_k, m_{k-1})\right] \\
&\quad + \left[F(u, u) - F(u, m_{n-1})\right] \quad \textbf{up}
\end{aligned}
$$

(3)

As an addition to these formulas, a program must now be implemented, that stores and deletes the local minima and maxima at the right time.

It remains to show that the simulation and the measurement agree. A total of 55 measured values from figure 1 were stored for this purpose. So 10 measurements were saved from the largest loop, 9 from the second largest, etc., and finally only one reading was saved from the smallest. 55 measured values appears very little, which makes the results achieved all the more remarkable. Figure 10 shows the simulated hysteresis curve which results from the input voltage profile from figure 1. Since it is a decaying sine wave, local minima and maxima are stored.

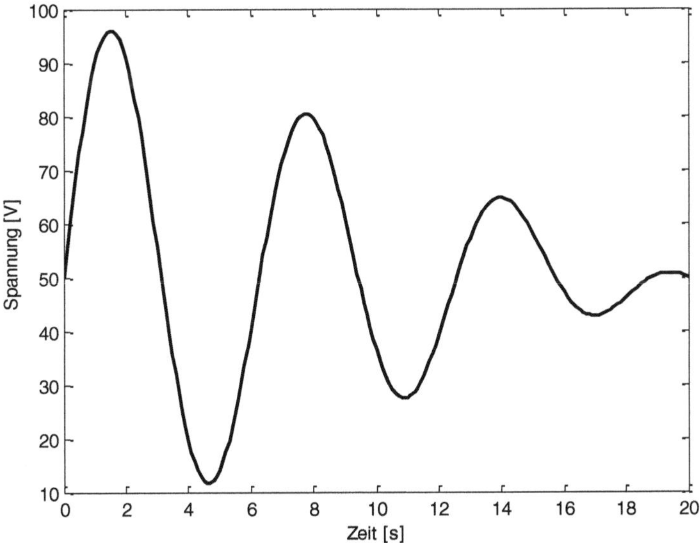

Fig. 9: Input signal u(t)

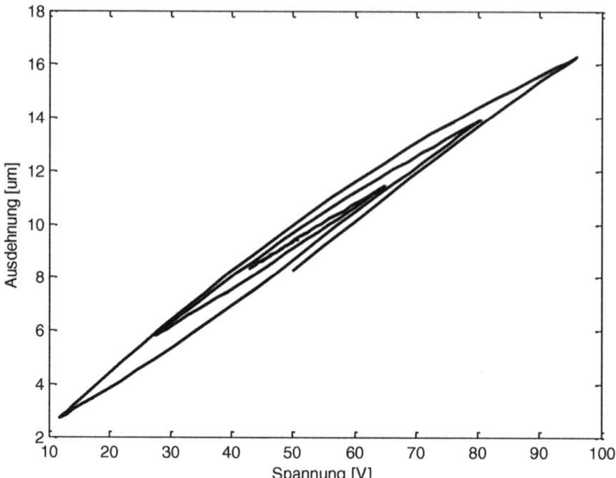

Fig. 10: Simulation

6.   Machine Learning approach

A method from the field of artificial intelligence is used as a further approach for calculating the weight function. First of all, any area function is assumed as the starting value. In the source code this is the matrix f_Learn. If one were to derive this twice according to formula (4) (same as 2), one would arrive at a homogeneous weight function μ(α,ß).

$$\mu(\alpha,\beta) = \frac{d^2 F(\alpha,\beta)}{d\alpha\, d\beta} \qquad (4)$$

This means that the calculation can be carried out using an input signal according to figure 11 and the formulas according to chapter 5. The input signal was chosen as is because several local maxima and minima must be stored when it is used for the hysteresis calculation. With these starting values, the calculation will certainly not match the measurement, which is also shown in figure 12.

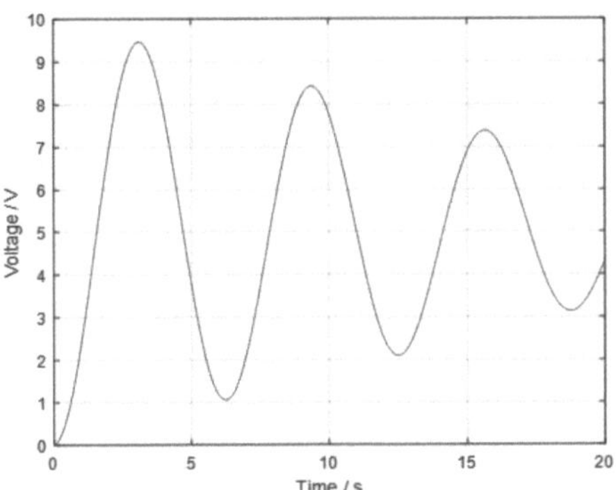

Fig. 11: Input signal u(t)

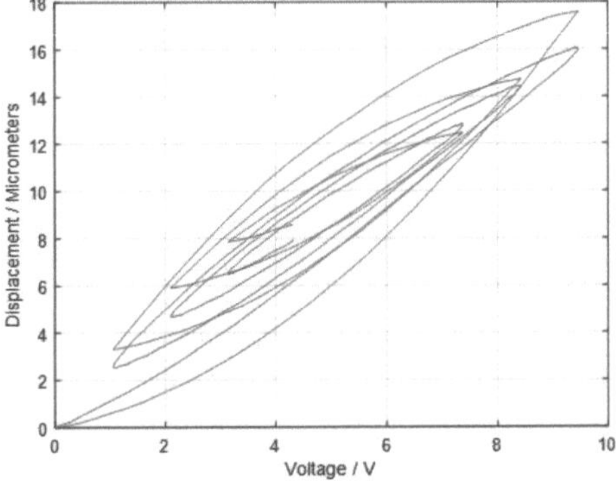

Fig. 12: Comparison of the untrained system with the measurement

In the core algorithm, the area function (f_learn in the code) is slightly modified parameter by parameter by adding or subtracting a small number with a random sign. Then the calculated hysteresis is compared to the measured one and the error is squared. If the total error after the calculation is smaller than before, the modified parameter is left as it is, otherwise it is replaced with the original value again. The calculated hysteresis gradually approaches the measured one. The result after 1000 iterations of all parameters is shown in figure 13.

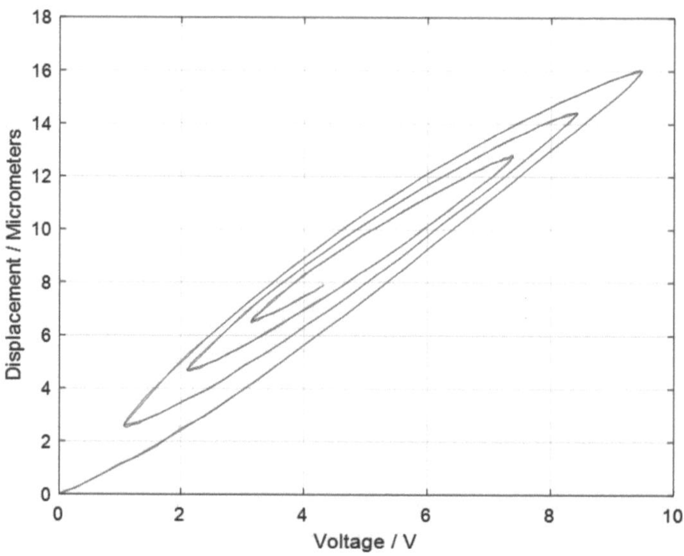

Fig. 13: Comparison of the measurement with the calculation trained with 1000 loops

7. Discussion and Matlab code

It turns out that usable results can be achieved with very few measured values or with a few learning loops. Once the weight function has been determined, the algorithm for calculating the function values can be easily implemented in the (signal) processors connected upstream of the actuators. However, one must be aware that storing local maxima and minima can lead to overflows very quickly. The decaying sine example above is the worst case; without ever erasing maxima and minima, more and more such are generated. This circumstance must be taken into account with suitable termination criteria.

Matlab Codes:

```
%piezohysteresis
clear;
f = [0 2.38 4.46 6.4  8.29 10.09 11.69 13.18 14.53 15.74 16.90;
0 2.32 4.36 6.30 8.18 9.86 11.40 12.88 14.13 15.36 16.52;
0 2.17 4.23 6.10 7.86 9.49 10.99 12.36 13.68 14.91 0;
0 2.15 4.11 5.90 7.64 9.19 10.56 11.91 13.23 0  0;
0 2.03 3.97 5.74 7.34 8.71 10.11 11.46 0 0 0;
0 2.01 3.76 5.39 6.84 8.23 9.63 0 0 0 0;
0 1.92 3.54 5.05 6.41 7.8 0 0 0 0 0;
0 1.73 3.22 4.59 5.95 0 0 0 0 0 0;
0 1.62 2.94 4.31 0 0 0 0 0 0 0;
0 1.43 2.75 0 0 0 0 0 0 0 0;
0 1.43 0 0 0 0 0 0 0 0 0]

m(1)= 0.0001;
M(2)= 0.0001;
m(2)= 0.0001;
n = 2
up = 1;
t = 0:0.1:20;
for k = 1:201
y (k)  = 5 + 4.99*(1-t(k)/20)*sin(t(k));
end
```

```matlab
k=1;
u = y(k);
for k = 1:201
    ualt = u;
    u = y(k);
    if up == 1
        if ualt > u
            M(n+1) = ualt;
            up = 0;
        else
            if (n>2) & (u>=M(n))
                n=n-1;
            end
        end
    else
        if ualt < u
            m(n+1) = ualt;
            n=n+1;
            up = 1;
        else
            if (n>2) & (u<=m(n))
                n = n - 1;
            end
        end
    end

out(k) = hysteresis(u,up,n,M,m,f);

end
plot(y,out);


function y = smoothmatrix(f,n,A,B)
b = B+1;
a = n-A;          % right numeration

bi = fix(b); br = b-bi;
ai = fix(a); ar = a-ai;         %Integer and Rest

r = f(ai,bi)+(f(ai,bi+1)-f(ai,bi))*br;
q = f(ai+1,bi)+(f(ai+1,bi+1)-f(ai+1,bi))*br;
y = ar*q+(1-ar)*r;

end


function x = hysteresis(u,up,n,M,m,f);
if up == 1
    x = 0;
    for k = 1:n-1
     x      =      x      +      (smoothmatrix(f,11,M(k+1),m(k+1))-
smoothmatrix(f,11,M(k+1),m(k)));
```

```matlab
    end
    x       =       x       +       (smoothmatrix(f,11,u,u)       -
smoothmatrix(f,11,u,m(n))));
else
    x = 0;
    for k = 1:n-1
        x       =       x       +       (smoothmatrix(f,11,M(k+1),m(k+1))-
smoothmatrix(f,11,M(k+1),m(k))));
    end
    x       =       x       +       (smoothmatrix(f,11,M(n+1),u)       -
smoothmatrix(f,11,M(n+1),m(n))));
end
end
```

Matlab Code, Machine Learning Approach

```matlab
%Hysteresis
clear;clf

% Initial_Values_Synapses
f_learn = [0 03.3 06.3 09.0 11.4 13.1 14.9 16.4 17.6 18.7 19.3;
0 03.0 05.7 08.1 10.2 12.0 13.5 14.7 15.6 16.2 16.5;
0 02.7 05.1 07.2 09.0 10.5 11.7 12.6 13.2 13.5 0;
0 02.4 04.5 06.3 07.8 09.0 09.9 10.5 10.8 0 0;
0 02.1 03.9 05.4 06.6 07.5 08.1 08.4 0 0 0;
0 01.8 03.3 04.5 05.4 06.0 06.3 0 0 0 0;
0 01.5 02.7 03.6 04.2 04.5 0 0 0 0 0;
0 01.2 02.1 02.7 03.0 0 0 0 0 0 0;
0 00.9 01.5 01.8 0 0 0 0 0 0 0;
0 00.6 00.9 0 0 0 0 0 0 0 0;
0 00.3 0 0 0 0 0 0 0 0 0];

% Testing_signal
t = 0:0.1:20;
for k = 1:201
y (k) = 5+4.99*(1-t(k)/30)*(-cos(t(k)));
end

% Initial Output
out_learn = hysteresisloopcalc (f_learn,y);

% Measurement Values
out = [0.0058  0.0638   0.1434   0.2422   0.4836   0.7479   1.0731   1.4272 ...
       1.8837   2.3951   2.9823   3.6672   4.3389   5.1523   5.9653   6.7624 ...
       7.6158   8.5229   9.3855  10.2176  11.0180  11.7373  12.4919  13.1545 ...
      13.7107  14.2835  14.7121  15.1070  15.4258  15.7129  15.8898  15.9971 ...
      16.0389  16.0223  15.9114  15.7891  15.6159  15.3903  15.1169  14.8399 ...
      14.4366  14.0677  13.5817  13.0785  12.5265  11.9783  11.3147  10.7122 ...
```

```
   10.0317    9.2937    8.6234    7.8646    7.1992    6.5137    5.8611    5.2127 ...
    4.6477    4.1325    3.7186    3.2937    2.9827    2.7492    2.5952    2.5605 ...
    2.5114    2.6006    2.6636    2.8572    3.0186    3.3050    3.6054    3.9274 ...
    4.3383    4.8299    5.2897    5.8247    6.4284    6.9866    7.6042    8.2809 ...
    8.8953    9.5506   10.2155   10.8026   11.3153   11.8377   12.3284   12.7799 ...
   13.1869   13.5410   13.8693   14.0591   14.2739   14.3546   14.4219   14.4401 ...
   14.4129   14.3829   14.2321   14.0815   13.9291   13.6590   13.4347   13.1397 ...
   12.7665   12.3970   12.0005   11.6205   11.1720   10.7069   10.1929    9.7369 ...
    9.1953    8.7389    8.1980    7.7432    7.2675    6.7797    6.3578    5.9705 ...
    5.6250    5.3260    5.1169    4.9201    4.7373    4.6894    4.6535    4.6635 ...
    4.7155    4.7671    4.9361    5.0605    5.2979    5.4859    5.7503    6.0884 ...
    6.4135    6.7604    7.0900    7.4831    7.8904    8.3450    8.7270    9.1520 ...
    9.5723   10.0219   10.3762   10.7510   11.1424   11.4671   11.7620   11.9844 ...
   12.2113   12.4016   12.5540   12.7077   12.7421   12.8175   12.8177   12.7449 ...
   12.7209   12.5873   12.5055   12.3154   12.1790   11.9392   11.7185   11.4797 ...
   11.2254   10.9926   10.7067   10.3722   10.0730    9.8091    9.4523    9.1336 ...
    8.8202    8.5571    8.2272    7.9416    7.6707    7.4609    7.2363    6.9996 ...
    6.8329    6.7377    6.6349    6.5250    6.4882    6.4812    6.5415    6.5473 ...
    6.6574    6.7501    6.8242    6.9580    7.1096    7.3173    7.4997    7.6617 ...
    7.8975]

% Learning algorithm
for loop = 1:1000;

for za = 1:11   %loop over the neurons
  for zb = 2:11
     %out = hysteresisloopcalc (f,y);
     out_learn = hysteresisloopcalc (f_learn,y);
     ls = leastsquare(out,out_learn);
     sign = fix(2*rand(1))*2-1; %random sign for change
synapse;
     f_learn (za,zb) = f_learn (za,zb) + sign*0.005; %learning
delta
     out_learn_new = hysteresisloopcalc (f_learn,y);
     ls_new = leastsquare(out,out_learn_new);   %Backpropagation
     if (ls_new >= ls)
       f_learn (za,zb) = f_learn (za,zb) - sign*0.005;
     end
  end
end
end
ls
loop
end

plot(y,out); hold on
plot(y,out_learn); grid
xlabel('Voltage / V')
ylabel('Displacement / Micrometers')

function y = leastsquare(a,b)
y = 0;
```

```matlab
for k = 1:length(a);
    y = y + (a(k)-b(k))^2;
end
end

function y = interpolated_matrix(f,n,A,B) %linear value
interpolation
b = B+1;
a = n-A;          % rigt numbering

bi = fix(b); br = b-bi;
ai = fix(a); ar = a-ai;          %Separation in Integer und Rest

r = f(ai,bi)+(f(ai,bi+1)-f(ai,bi))*br;
q = f(ai+1,bi)+(f(ai+1,bi+1)-f(ai+1,bi))*br;
y = ar*q+(1-ar)*r;
end


function x = hysteresis(u,up,n,M,m,f);
if up == 1
    x = 0;
    for k = 1:n-1
     x = x + (interpolated_matrix(f,11,M(k+1),m(k+1))-
interpolated_matrix(f,11,M(k+1),m(k)));
    end
   x = x + (interpolated_matrix(f,11,u,u) -
interpolated_matrix(f,11,u,m(n)));
else
    x = 0;
    for k = 1:n-1
     x = x + (interpolated_matrix(f,11,M(k+1),m(k+1))-
interpolated_matrix(f,11,M(k+1),m(k)));
    end
   x = x + (interpolated_matrix(f,11,M(n+1),u) -
interpolated_matrix(f,11,M(n+1),m(n)));
end
end

function output = hysteresisloopcalc(f,y)
m(1)= 0.0001;
M(2)= 0.0001;
m(2)= 0.0001;
n = 2;
up = 1;
k=1;
```

```
u = y(k);
for k = 1:201
    ualt = u;
    u = y(k);
    if up == 1
        if ualt > u
            M(n+1) = ualt;
            up = 0;
        else
            if (n>2) & (u>=M(n))
                n=n-1;
            end
        end
    else
        if ualt < u
            m(n+1) = ualt;
            n=n+1;
            up = 1;
        else
            if (n>2) & (u<=m(n))
                n = n - 1;
            end
        end
    end
output(k) = hysteresis(u,up,n,M,m,f);

end
end % of hysteresisloopcalc
```

8. Literature

[1] Peng, J. Y., & Chen, X. B. 2012. Novel Models for One-sided Hysteresis of Piezoelectric Actuators. *Mechatronics.* 22(6): 757–765.

[2] Gu, G., & Zhu, L. 2010. High-speed Tracking Control of Piezoelectric Actuators Using an Ellipse-based Hysteresis Model. *Review of Scientific Instruments.* 81(8)

[3] Codourey, Alain, et al. "High precision robots for automated handling of micro objects." *Seminar on Handling and Assembly of Microparts, Vienna, Austria.* 1994.

[4] Zesch, Wolfgang, et al. "Inertial mechanisms for positioning microobjects: Two novel mechanisms." *SPIE Photonics East'95: Microrobotics and Micromachanical Systems Symposium, Philadelphia PA, USA, 1995.* Vol. 2593. SPIE, 1995

[5] Büchi, R. (1996). Modellierung und Regelung von Impact Drives für Positionierungen im Nanometerbereich (Doctoral dissertation, ETH Zurich).

[6] Mayergoyz, I. D, Mathematical Models of Hysteresis, 1994, ISBN 0-387-97352-4

[7] Preisach, F. Über die Magnetische Nachwirkung, Z. Phys 94, p 277, 1935

[8] Russel, Stuart J, Norvig, Peter, Artificial Intelligence: A Modern Approach (2nd edition), Upper Saddle River, New Jersey: Prentice Hall, pp 111- 114, ISBN 0-13-790395-2, 2003.

[9] Unbehauen H. Regelungstechnik. Braunschweig: Vieweg; 1992.

[10] Büchi, Roland. *State space control, LQR and observer: step by step introduction with Matlab examples*. Norderstedt Books on Demand, 2010.

[11] Dimmler, Martin, Ulf Holmberg, and Roland Longchamp. "Hysteresis compensation of piezo actuators." *1999 European Control Conference (ECC)*. IEEE, 1999.

[12] Lin, Chih-Jer, and Sheng-Ren Yang. "Precise positioning of piezo-actuated stages using hysteresis-observer based control." *Mechatronics* 16.7 (2006): 417-426.

[13] Sabarianand, D. V., P. Karthikeyan, and T. Muthuramalingam. "A review on control strategies for compensation of hysteresis and creep on piezoelectric actuators based micro systems." *Mechanical Systems and Signal Processing* 140 (2020): 106634.

[14] Buchi, Roland, et al. "A remote controlled mobile mini robot." *MHS'96 Proceedings of the Seventh International Symposium on Micro Machine and Human Science*. IEEE, 1996.

[15] Ayala, Helon V. Hultmann, et al. "Nonlinear black-box system identification through neural networks of a hysteretic piezoelectric robotic micromanipulator." *IFAC-PapersOnLine* 48.28 (2015): 409-414.

[16] Somov, Ye I. "Modelling physical hysteresis and control of a fine piezo-drive." *2003 International Conference Physics and Control. Proceedings*. Vol. 4. IEEE, 2003.